ENERGY

THE POWER OF CANADA

ENERGY

THE POWER OF CANADA

Fitzhenry & Whiteside

Energy: The Power of Canada

© Minister of Supply and Services Canada 1988
Catalogue no. M22-108/1988E

Published by Fitzhenry & Whiteside Limited in cooperation with Energy, Mines and Resources Canada and the Canadian Government Publishing Centre, Supply and Services Canada.

Design: Addventures/Ottawa
Graphics Editor: Debora Morton, Addventures/Ottawa
Typesetting: Graphcomp Design/Ottawa
Film Assembly: MBD Litho Plate Inc./Markham
Colour separations: Olympic Scanning Inc./Scarborough
Printing: Ashton-Potter/Concord
Binding: Anstey Graphics Ltd./Toronto

Canadian Cataloguing in Publication Data

1. Main entry under title:

Energy: The Power of Canada

Co-published by Energy, Mines and Resources Canada
Issued also in French under title: *Énergie: l'horizon canadien*
ISBN 0-88902-106-6

1. Power, Resources, Canada.   I. Canada. Energy, Mines and Resources Canada.

TJ163.25.C3E63 1989   333.79′0971   c88-095323-3

PRINTED AND BOUND IN CANADA

SHAPING CANADA'S DESTINY

It would be impossible to understand Canada or imagine its future without knowing something about energy. The chronological, constitutional and to some extent emotional history of this country has its parallel in the exploitation and discovery of different kinds of fuel.

The passion with which Canadians have characteristically debated issues such as the ownership, taxation, development and transportation of energy attests to its importance. Although these debates have sometimes led to the inflammation of regional tensions and struggles for power between different levels of government, they are above all evidence of a strong and growing country in the process of coming to terms with the geographical and economic forces that help to shape it.

We are on the verge of entering a new century. The world is facing the challenge of trying to ensure universal access to adequate supplies of energy from a variety of sources without worsening the environmental crisis already upon us. That is why the decisions we are taking now about energy will have such profound implications for the future.

It is within our capacity to bequeath to future generations a clean, safe world. This book is intended to encourage Canadians to think about energy not as an obstacle to that goal but as a means to achieving it.

The Honourable Marcel Masse
Minister of Energy, Mines and Resources

TOMORROW'S RESOURCES

We are coming into a time in which we will rely increasingly on a variety of fuels to meet our energy needs and in which respect for the environment will be a critical component of energy development. Over the last decade Canadians have become increasingly aware of the relationship that exists between energy use, economic growth and environmental degradation.

In Canada we have already begun to establish the infrastructure for a multiple-energy economy. The fact that gains from energy efficiency and conservation can be counted as energy supplies in the same way as coal or petroleum reserves and that energy-efficient economies are also better able to compete internationally forces us to change our concept of what constitutes a resource.

This book hopes to expand the energy horizons of Canadians and to inspire them to become leaders in environmentally sustainable economic development.

FORGING NEW ENERGY HORIZONS

Canada is a nation blessed with a wide variety of energy resources. Our country has abundant supplies of bitumen, natural gas, coal, uranium, and hydroelectric power. Biomass, which includes forestry and wood processing residues, crop remains and animal wastes, is a versatile and plentiful fuel source whose potential has barely been tapped.

In 1987, industrial market-oriented countries saved $250 billion on expenditures for oil, gas, coal and nuclear power because of gains in energy efficiency. Although most Canadians are not accustomed to thinking this way, gains from energy efficiency and conservation still to be achieved are calculated as actual energy resources. It has been estimated that Canada could realistically cut its energy consumption by another 20 percent over the next decade.

Despite the sense that with lower prices the energy "crisis" of the 1970s is over, it would be unwise to adopt an attitude of complacency toward the future. The devastating rise in oil prices that shattered presumptions about stability of petroleum supplies and led many Canadians to question for the first time whether we were making profligate use of non-renewable resources may never occur again, but there are other important reasons why we should continue to pursue energy efficiency and conservation in the future.

Over the last decade a number of reports, including those of the World Commission on Environment and Development and of the Canadian Energy Options group, have tried to focus public attention on the two concerns that will dominate energy policy-making in the future.

The first involves the necessary transition that industrial economies are undergoing from a reliance predominantly on oil to a dependence on different kinds of energy and on energy efficiency and conservation. As Japan has proven, improved international productivity and competitiveness are related to investment in energy-saving technology.

Canada has a head start on the rest of the world where multiple energies are concerned. Not only are we endowed by nature with a number of options, but Canadians have also been in the forefront of developing the technologies and infrastructures that make it possible to provide secure and affordable energy from a variety of sources. The transition to a multiple-energy economy, while inevitable, is by no means complete.

The second is the relationship between energy use, the economy and the environment. After the price shocks of the 1970s many Canadians began to drive smaller cars and to try to conserve energy more efficiently in their homes. Wood stoves became fashionable and many oil furnaces were replaced by ones fired by natural gas or electricity. A few thousand innovative Canadians built homes heated by solar panels, purchased solar water heaters or converted their car engines to run on propane fuels.

Industries are beginning to realize that the cost of environmental protection is part of the responsibility of being in business.

At first these changes were inspired less by concern about damage to the environment caused by the burning of fossil fuels than by concern about damage to our pocketbooks. However, the incidental increase in our awareness of the environmental crisis we now face has meant that even with the radical drop in oil prices, the search for better ways to conserve energy should be of concern to all Canadians.

Few Canadians are indifferent to the gradual environmental degradation caused by the exploitation and consumption of energy resources. Acid rain, the warming of the earth's climate, ozone depletion, water pollution — we become more aware of these by-products of industrialization every day.

The cost of environmental degradation has not conventionally been part of the price Canadians pay for energy and in the past environmental concerns were not always included in the energy policy-making process. This is now changing. To reduce acid rain, for example, Canada is committed to lowering emissions of sulphur dioxide by 50 percent by 1994. While installing scrubbers on smokestacks has been one traditional and expensive solution, researchers are now looking at ways of controlling pollutants at the combustion stage. In the meantime, fuel substitution, coal washing and blending, and burner modifications are helping to alleviate the problem significantly. According to Environment Canada, sulphur dioxide emissions decreased by 44 percent between 1970 and 1984.

The use of nuclear power to produce electricity does not contribute to acid rain or the greenhouse effect. Nuclear wastes, however, are highly toxic. Finding a method to dispose of these wastes safely is therefore a global concern.

Even the development of hydroelectric power can pose a threat to the environment. Falling water itself is a renewable resource and electricity is pollution free at the point of use but the large projects typically result in environmental disruption, such as the flooding of reservoirs and destruction of fish habitats.

Our survival will depend in the future on the compatibility we can achieve between our economic goals and environmental protection. In Canada, where energy policy involves not only the development of new technology but sensitivity to regional concerns and job creation, several levels of government and industry are now working to coordinate their efforts. The result should be continuing prosperity for Canadians that does not imperil the welfare of the earth's environment.

Canada's offshore resources are being developed through the cooperation of industry and government.

"I noticed in space a certain shift in perception — you begin to value things that are taken for granted on earth, like fresh air, a mountain stream, sunrise in the forest, a shivering of leaves on the trees. All these things we do not pay much attention to down here."

Georgi Grechko, Soviet cosmonaut.

OUR ENERGY ENVIRONMENT

Try to imagine a world without energy. Try to imagine a world without heat, without light. In Canada, with a land mass that spans 9.8 million square kilometres and a climate famous for its harsh winters, energy and the services it provides have been essential to survival since long before Confederation.

In our contemporary world, energy plays a major role in every aspect of our lives — in our homes, our workplaces, our industries and in transportation. Whether we cook with natural gas or electricity, whether we ride public transit or drive a car, whether we shop in air-conditioned super-markets or work in climate-controlled office towers, we are using energy to feed ourselves, heat our homes, transport us from place to place and create environments in which we can work and play. We have come to take the services energy provides so much for granted we spend little time thinking about the actual mechanics of how energy is transformed into the comforts of modern life.

Most of our energy is produced from crude oil, natural gas, coal and large hydroelectric facilities. About seven percent comes from renewable sources such as biomass, solar, small hydro and wind energy. Another seven percent is derived from nuclear power.

Energy in Canada is big business. The energy sector directly employs more than 300 000 people, representing almost three percent of total employment in Canada. In 1986, gross revenues for the sector topped $60 billion. In fact, Canada produces about four percent of the world's primary energy supply and among OECD (Organization for Economic Cooperation and Development) countries is second only to the United States as a producer of energy.

Canada is also a major exporter of energy. Over

80 percent of our energy exports go to the United States alone. Both Canadians and Americans are often surprised to learn that Canada is one of the biggest foreign suppliers to the U.S. of petroleum, natural gas and electricity.

Energy has also played an increasingly significant role in our international trade. Between 1970 and 1986, energy commodities increased from 6 to 10 percent of our total exports. In 1986 alone, energy exports grossed revenues of more than $12 billion.

While increasing production and exports, Canadians have made conscientious efforts to reduce energy consumption at home. Despite major improvements in energy efficiency, however, Canada remains one of the most voracious energy-consuming countries in the world. Since 1979, energy consumption in Canada has dropped 25 percent but experts estimate that potential gains of up to 20 percent remain to be met over the next decade.

Not surprisingly, industry is the most energy-hungry sector of our economy, responsible for some 37 percent of the country's energy demand. The transportation sector ranks second, consuming about 27 percent of the energy we use, the residential sector (the one with which Canadians as a whole are most familiar) consumes approximately 21 percent, and the commercial sector (our offices, schools, hospitals and government buildings), 15 percent.

CANADA'S ENERGY HERITAGE

Having evolved through an agricultural to an industrial economy, Canada is now in a slow transition to a post-industrial society. Over the last century Canadians have gone from a reliance on wood to coal and subsequently to oil, gas and electricity.

THE WOOD ERA

A little more than 100 years ago wood accounted for about 85 percent of the country's energy production and was the sole source of heat both for the hearth and the stove. Canada's abundant forests provided the aboriginal people and settlers alike with a seemingly endless and ready supply of fuel.

In our contemporary world, energy plays a major role in every aspect of our lives.

Of course, there were other sources of energy available at the time. Coal, for example, was already emerging as the preferred fuel for industrial processes. Wind propelled sailing ships and powered windmills. Water power was used in the logging industry and animal power was the major source of energy in the agricultural sector.

Wood energy is again popular in areas of the country. In parts of the Atlantic provinces where supply is ample and conventional fuel prices high, cordwood can regularly be seen stacked alongside the house or against the back shed.

Bioenergy (energy from any plant or animal material, including wood) currently provides almost seven percent of Canada's total energy supply. It is experiencing a resurgence in popularity in the industrial as well as the residential sector. For years, the pulp and paper industry has provided much of its own energy requirements by burning spent pulping liquors and wood wastes. Now, some commercial and institutional buildings are burning wood chips to meet most of their space heating requirements. This is a trend which is likely to become more evident in the future.

Logging in B.C.: Bioenergy provides almost seven percent of Canada's energy needs.

THE COAL ERA

The Industrial Revolution ushered in the era of coal, a combustible sedimentary rock. Coal burns hotter than wood. It contains more energy and is easier to transport. The introduction of the now classic steam engine and factory processes requiring high temperatures demanded just such a fuel.

Coal had been mined in Canada as early as 1639, when a small colliery was opened in Grand Lake, New Brunswick, to supply the fortress at Louisbourg. By the turn of the century it had surpassed wood as the most dominant fuel and as late as 1945 was supplying over half of Canada's overall energy needs.

Although coal was plentiful, it was eclipsed by oil and natural gas in the early 1950s. Today, coal accounts for about 14 percent of Canada's energy supply and produces 16 percent of Canada's electricity. Alberta, Saskatchewan and Nova Scotia are almost completely dependent on coal for electricity generation.

Canada has such abundant coal reserves (nearly four percent of the world's supply) that at current rates of consumption the country's requirements could be met for several centuries. In fact, the energy potential of Canada's coal far exceeds that contained in all of this country's oil, natural gas and oil sands combined.

The historical difficulty with coal, as with oil and gas and hydroelectric power, is that nature did not lay down Canadian resources close to Canadian markets. Traditionally, policies dealing with natural resources have involved devising compromises between the logic of a north-south market axis and the willed reality of an east-west nation.

Nevertheless, today Ontario Hydro buys about 34 percent of its coal from Western Canada because coal from that area of North America is low in sulphur content and thus helps Ontario meet its goals for reduction of acid gas emissions.

In 1987, Canada produced 61 million tonnes of coal, four million more than in 1986. Of this, about 27 million tonnes were exported, mainly to Pacific Rim countries (it is our largest export commodity to Japan and South Korea), but with significant sales to Europe and South America as well. As a result, coal, which is exported primarily for use not as a

Coal destined for central Canada is off-loaded from a Canadian collier.

18

fuel but as a component in the metallurgical manufacturing process, is, after oil and gas, Canada's third most valuable mineral export commodity. Its annual sales of $2 billion are roughly equal to the value of Canada's Japanese car imports.

The coal industry currently provides direct employment for nearly 12 000 Canadians and indirectly employs another 30 000. An additional 200 000 jobs are found in coal-related industries, including the steel, transportation, utilities, service and consumer sectors.

Many Canadians think of coal as an old-fashioned,

Canada's oil industry dates back to the 1947 discovery at Leduc, Alberta.

somewhat undesirable energy resource. It is associated in their minds with unhealthy working conditions in dark underground mines or, more recently, with environmental pollution, especially acid rain. The search for environmentally acceptable ways to burn coal is now taken seriously by industry and government. As the demand for this fuel by Canada's electricity generating plants is expected to double over the next decade, utilities interested in the clean and economic use of coal may soon have a variety of new technologies, such as circulating fluid beds and integrated combined cycle systems, from which to choose. These technologies will cost less, be more efficient and offer better environmental control for both new and existing plants than scrubber systems now being used.

THE OIL ERA

Over the last 40 years, oil and gas have emerged as dominant energy sources in Canada and around the world. Oil was easier to transport than wood or coal. In addition, oil could be refined into a variety of products for industrial use. Most importantly, it was a convenient transportation fuel, which quickly encouraged the widespread use of gasoline and diesel engines.

In Canada, as elsewhere, the transition to oil was sudden and dramatic. Almost overnight, the petroleum industry established a sound economic base, particularly in Western Canada. Later discoveries have created the prospect of future economic benefits in the Arctic and off the Atlantic coast.

The discovery of oil transformed the Canadian economy, and, eventually, the relationship between the provinces with oil and those without, and that between Western Canada and the government in Ottawa. The problems created by the discovery of oil and gas far from hungry Canadian markets paralleled to some extent the experience of the coal industry. The difference was the transportability of petroleum. The construction of pipelines to carry Western Canadian oil to Ontario, but no further, was another in the series of attempts to find a compromise between the geography and market structure of North America. When the price of imported oil rose suddenly in 1973, the vulnerability of this policy was exposed. The debates that raged across Canada during the 1970s were frequently about energy and its role in building a nation. Because of the immense size of Canada and its particular history, issues that arise under energy policy have always been, and probably always will be, some of the most controversial facing the country.

Today, oil remains Canada's largest primary energy source, still strongly dominating certain sectors of our economy, particularly transportation. Its contribution to our overall energy use, however, has dropped from 51 to about 36 percent, making Canada one of the least oil-dependent industrial economies in the world. This is the result of the success of energy-efficiency improvements and conversions from oil to natural gas, electricity, wood and other renewable sources of energy.

Perhaps it is because Canada claims such an abundance of petroleum resources — we are the world's third largest producer of natural gas and the ninth largest producer of crude oil — that Canadians tend to take energy for granted. Despite their short histories, we refer to oil and gas as conventional fuels. In fact, Canada's oil and gas industry can trace its beginnings only as far back as 1947, to the

To date 4.2 trillion cubic metres of natural gas have been discovered in Canada.

22

The oil and natural gas industry is serviced by an extensive transportation and refining system.

discovery of the Leduc oil field in Alberta. By the end of 1987, 168 000 wells had been drilled in Canada.

The impact of the oil industry on our economy has been tremendous. In 1987, the value of output from Canada's fuel sector was about $20 billion. Of this, petroleum production represented $12 billion and natural gas production, $4.3 billion. Crude oil production averaged 254 000 cubic metres per day, with 159 000 cubic metres of this destined for domestic demand and most of the remainder exported to the United States. Canadian domestic natural gas sales in 1987 amounted to 47.8 million cubic metres; another 28 million cubic metres were exported south of the border.

The oil and natural gas industry in Canada is serviced by an extensive transportation and refining system including over 35 000 kilometres of crude oil pipelines and 172 000 kilometres of natural gas gathering systems and pipelines. To date, three billion cubic metres of conventional oil and 4.2 trillion cubic metres of natural gas have been discovered in Canada. More important still is the enormous undiscovered potential of the Western Sedimentary Basin — an area of approximately 1.5 million square kilometres extending from British Columbia eastward through Alberta, Saskatchewan and Manitoba. It is estimated that 636 million cubic metres of conventional oil and 2.5 trillion cubic metres of natural gas remain to be discovered.

In addition, Canada has vast deposits of heavy oil and oil sands in northern Alberta and Saskatchewan. Canada's oil-sands deposits, the largest in the world, are estimated to contain the equivalent of the combined proven oil reserves of Saudi Arabia, Kuwait and the United Arab Emirates. These deposits, however, are considerably more expensive to develop than conventional reserves. Intensive research and development are being carried out to reduce costs and improve our technical understanding of oil-sands recovery and the production of synthetic light crude oil through upgrading.

THE MULTIPLE FUELS ERA

We have already witnessed a decline in the use of wood, and over the next 20 years oil and gas will gradually be replaced by other fuels. By the mid-1990s, for example, Ontario will be deriving 60 percent of its power needs from nuclear generating plants. These changes do not so much reflect the merits of the various fuels as they do the changing demands of our society.

Nevertheless, wood, coal, oil and gas will continue to play a role in energy use. Part of that role will be as generators of electricity. Electricity can be produced from many different sources — from water, from oil, gas and coal, from nuclear fission, wind, biomass and solar. As the world's fourth largest producer of electricity after the United States, Japan and the Soviet Union, and the second most intensive user of electricity after Norway, Canada has developed world class technologies in all aspects of the production, distribution and utilization of electric power.

Canada has, for example, supported the development

of its own indigenous nuclear power technology — the CANDU. These reactors are used in New Brunswick, Quebec and Ontario. They are acknowledged universally as the safest and most sophisticated in the world.

It is a much older energy source — hydro power — that has been our most important generator of electricity. In fact, Canada is the world's largest producer of hydro-electricity, generating almost 68 percent of its electric power from falling water.

All of the fuel used to power CANDU nuclear reactors is processed in Canada.

Construction work at the Darlington nuclear plant in Ontario.

During the 1940s and '50s, much of this country's hydro power came from hundreds of small, dispersed sites. Many of these were subsequently abandoned in favour of large central facilities. Today, Canada depends on large-scale projects for most of its hydroelectricity. James Bay in Quebec, the Nelson River in Manitoba, the Peace River in British Columbia and the Churchill Falls development in Newfoundland — all produce massive quantities of electricity from water.

In today's competitive and variable energy marketplace, however, private sector entrepreneurs, municipalities and many provincial governments are encouraging a return to smaller, less financially and environmentally risky hydro projects.

In addition to small hydro, there are other new and renewable sources of energy which may contribute further toward satisfying our thirst for electricity. These include cogeneration, solar electricity and wind. In cogeneration, one fuel source (it may be natural gas, biomass, coal, municipal waste or any other fuel) is used to produce two energy products simultaneously — heat and electricity. Many industries that already burn fuel to produce heat for

Wind turbine at Atlantic
Wind Test Site, P.E.I.

thermal processes can, by installing a cogeneration unit, generate electricity for essentially the same fuel costs. This not only increases fuel efficiency, generates new revenue, and makes Canadian industry more internationally competitive, it also offers environmental benefits through lowered overall energy use. With research into residential uses making headway, it is quite possible to imagine the day when homeowners will be able to install their own mini-cogeneration units.

Solar and wind energy also offer environmentally sensitive solutions to electricity production. While the cost of solar electric (photovoltaic) systems is still high and the applicability of wind turbines is limited to parts of the country, these new technologies are becoming more popular in rural, remote and northern areas. The Canadian Coast Guard has already installed more than 3 000 small photovoltaic units across the country to power navigational beacons. For cost and reliability, they are a better choice than traditional battery systems.

In southern Alberta, modern wind turbines are becoming a more familiar sight for water pumping and electricity generation. Pioneers like farmer Ernie Sinnott, who installed the first privately owned large wind machine in Canada, are starting to change the way electricity is generated on the Prairies.

By looking at our past, we have attempted to offer some perspective on our future and provide a preview of things to come in the multiple fuels era. Whatever happens, it promises exciting advances in technology. Those advances, in combination with the drive for stable economic growth and a greater sensitivity to the fragility of the natural environment, should enable Canadians to create a safe and prosperous future.

Hydro-lines; Canada generates almost 68 percent of its electrical power from falling water.

"One must care about a world one will
not see."

Bertrand Russell (1872 – 1970)

OUR DAILY LIVES

Most Canadians are not interested in energy *per se*. As long as the car performs, the furnace provides heat, and the lights come on, we don't often think about the means by which these events occur.

The fact that the conveniences provided by energy have been available for only a few decades tends to be forgotten in the discussion of the past and the future. The enormous growth of the North American industrial economy in the 1950s would have been impossible without readily accessible, inexpensive fuel. In Canada, the belief that this situation, though entirely recent, was also eternal, was reflected in what we now think of as profligate use of energy in every sector of the economy and in the confident projections of industry and government about how much oil and gas there actually was.

The sense of complacency about energy and the wasteful consumption habits that took root during the era of cheap petroleum came to an abrupt end with the 1973 oil embargo. The fear that oil and gas supplies might run out changed the way Canadians thought and acted about energy. The shock of rising oil prices hit us right in the pocketbook. In the drive to conserve energy, Canadians discovered at the same time the intrinsic benefits of energy efficiency and, to our future benefit, began to develop new energy sources as alternatives for the years ahead.

During the same period, the link between energy use and pollution was becoming more evident to Canadians, as it was to millions of other people. Today it is increasingly understood that if we are to enjoy continued economic growth we must find ways to harmonize the development of energy resources with the limits of the natural environment. The successful design and application of technologies

that help achieve that goal depend as much on political decisions as on scientific research and individual choices.

OUR HOMES:
ENERGY LIFESTYLES

More than 80 percent of the energy we consume in our homes is for simple requirements such as space and water heating — services that can be provided by a variety of energy forms.

In recent years, some of the most outstanding advances in improved energy use have been made in the field of housing. The oil crisis sparked an awareness that our houses were full of costly drafts and air leaks. The battle against winter, previously fought with energy supply, can now be fought instead with energy efficiency.

With the support of government incentive programs, thousands of Canadians insulated and draft-proofed their houses and gave up their traditional oil furnaces to install new, improved natural gas, electric or wood-heating systems.

The battle against winter, previously fought with energy supply, can now be fought with energy efficiency.

Other Canadians took advantage of those incentives to purchase solar domestic hot water systems, which can provide up to 70 percent of average daily household water heating requirements. Solar water heaters have also been installed in new subdivisions and on apartment buildings. Energy-saving measures such as these helped some Canadians reduce their hot water bills by 50 percent or more.

Between 1973 and 1986, energy consumption per Canadian household declined from an equivalent of 26 barrels of oil annually to about 21 barrels. This was attributable both to the retrofitting of existing homes and to improvements made in the construction of new houses. Energy, Mines and Resources Canada reports that new houses today require, on average, 30 percent less energy to heat the same space than those constructed in 1974.

While they upgraded existing homes, Canadians learned many valuable lessons, which were applied to new housing. The result has been a new generation of houses that are bright, airy and weather-tight. Canadians have discovered the comfort and merits of extra insulation, controlled ventilation, passive solar heating, high-performance windows and a host of other innovations.

Over the last few years, Canadian home builders have come to think of a house as a total system integrated into its environment.

THE R-2000 HOME

The "house as a system" concept has resulted in a number of creative ideas that have led to more comfortable, more efficient dwellings. Almost as a by-product, in some cases, energy consumption has dropped. In other instances, energy efficiency has been intentionally designed into the home.

One example is the R-2000 house. A joint effort of Energy, Mines and Resources Canada and the Canadian Home Builders' Association, R-2000 ("R" for resistance to heat loss, and "2000" for the future) is the leading energy-efficient home technology in the world. More than a decade of research by government, private industry and universities has gone into the development of this technology.

THE SMART HOUSE

The "smart" house is another innovation that could change the way Canadians live. It is equipped with a computerized electrical system that replaces the usual tangle of wires running like spaghetti through the walls and basement of most homes. Instead, a single cable carries power for everything from the telephone to the toaster.

The first smart houses could be on the market in Canada by the mid-1990s. This is what is possible in one of the more sophisticated models: lights would automatically switch off when you leave the room; room temperatures would slowly drop during the night and then rise again in time for you to get out of bed; an alarm would sound to indicate any source of major heat loss; and you could program your kitchen appliances from your office microcomputer.

R-2000 home — high-performance windows allow sunlight to enter but prevent heat from escaping.

ENERGY-EFFICIENT APPLIANCES

Canadian companies are also developing and marketing everyday appliances, including microwaves, freezers and refrigerators, that use a fraction of the energy their older counterparts required.

WINDOWS

Windows are a major source of heat loss from both homes and commercial buildings. But now, more energy-efficient windows are reducing this heat loss, eliminating drafts and creating new business opportunities for the Canadian window industry.

High-performance windows are already commercially available. A transparent "low emissivity", or "low-e", coating on the window allows sunlight to enter a building but, like silver foil, prevents infra-red radiation (or heat) from escaping. For maximum effectiveness, this low-e coating is usually used in combination with a low conductive gas, such as argon, between the window panes. Heat loss can be further reduced by adding extra glazing to achieve triple- or even quadruple-layer windows.

HEATING WITH WOOD

As conventional energy prices rose in the 1970s, wood burning gave homeowners more control over energy costs. Scientific research, new installation and safety codes, and modern stove designs — these have altered the face of wood heating. Stoves are safer and more efficient and come in a wide variety of sizes and styles; new central wood-burning furnaces can heat an entire house evenly. As a result, wood is more popular than ever for supplementary heating.

RECYCLING

Across the country, a growing number of Canadian towns and cities have started recycling programs. This war on waste is not new, but the level of community interest and municipal responsibility is. Old newspapers, iron, glass, tires and aluminum are being collected, sometimes at central depots, sometimes by means of convenient door-to-door collection.

Why? Because recycling reduces pressure on landfill and generates local employment, and considerable energy savings. When an aluminum can is recycled, for example, 97 percent of the energy used in manufacturing it the first time is reclaimed.

In addition, all levels of government are investigating ways of burning waste (municipal garbage, landfill gas, sludge, wood residues, industrial and commercial solid wastes) in an environmentally sound manner as a method of waste disposal and a new source of energy supply. Not only would this provide municipalities with a new source of revenue, it would also reduce the amount of garbage that must be buried.

OUR BUILDINGS:
MEETING THE ENERGY CHALLENGE

It is not only the residential sector that has wrestled with the matter of energy efficiency. In the commercial and institutional sectors, millions of dollars are lost each day through energy waste that is preventable at reasonable cost.

Between 1965 and 1973, energy consumption in Canada's commercial sector rose at an average annual rate of 4.8 percent. Since then, it has fallen by about 27 percent. By designing new buildings and refurbishing existing ones for energy-efficient operation, by adding insulation, improving lighting, recovering heat and controlling ventilation, building owners are getting more out of the energy their properties consume.

Something as commonplace as lighting offers scope for remarkable energy savings. In hospitals, office towers, and large government and commercial buildings, conventional fluorescent lighting can easily account for as much as 40 to 60 percent of the total electricity bill. The installation of a simple electric current reducing device in the circuitry of fluorescent fixtures can effectively lower lighting costs by 35 percent.

As well as making good business sense, energy management in buildings reduces environmental problems,

both inside the building and out. Many energy-efficiency measures improve ventilation, lighting and air quality, while lowering emissions and waste products.

Where building mechanical systems are concerned, the possibilities for improvements in energy efficiency are almost limitless. Building owners and managers are doing the obvious: replacing old equipment and creatively integrating their heating, ventilation and air-conditioning requirements for more efficient operation. New technology includes high-efficiency boilers, air-sealing devices that prevent leakage, heat pump systems and high-performance windows. Some buildings are now heated entirely by "waste heat" from lights and machinery.

Given that residential and commercial buildings in

Since 1973, energy consumption in Canada's commercial sector has been cut by about 27 percent.

By using computer simulation programs, firms can analyze their manufacturing processes to determine where energy cost savings can be made.

Canada consume about 36 percent of the total energy load in Canada, building experts estimate that advanced construction technology and energy efficiency could reduce energy demand in Canada by 20 to 25 percent. The fact is that the benefits go beyond energy cost savings. New technology in this area is also helping Canadians raise their living standards, protect the environment, and increase our ability to compete in markets at home and abroad.

As well as making good business sense, energy management in buildings reduces environmental problems both inside and out.

"We must live by the world, and such as we find it, so we make use of it."

Michel de Montaigne (1533 – 1592)

FUELLING INDUSTRIAL GROWTH

Prior to the Arab oil embargo of 1973, industry, both in Canada and abroad, was essentially unconcerned about the security of energy supplies, and this attitude was reflected in industrial energy consumption and in transportation policies. It was also passed on to consumers, who were completely unprepared for the price shocks that transformed the economies of the industrialized countries.

By 1974, the world price of oil had risen from US$3 a barrel to US$12; by 1980, it was almost US$40. Industries worldwide became rigorous in their efforts to reduce energy-to-manufacturing cost ratios. The job of "plant energy manager", practically unheard of before the oil crisis, suddenly became commonplace.

By 1976, Canadian industry had recognized that energy efficiency could be an effective way to lower manufacturing costs and remain competitive in global markets.

In the process, many Canadian companies have dedicated personnel to the analysis and reduction of energy use in all company activities from the obvious (nighttime set-back of thermostats) to the obscure (installation of timers on parking lot electrical outlets).

Industry is looking for other ways to increase savings as well. Some are recovering waste products such as sawdust and wood trimmings or reclaiming "waste" heat from ventilation systems, air compressors, boilers and furnaces. Others have adopted new process designs, tightened manufacturing schedules, increased the use of robotics, introduced computer-integrated manufacturing systems, or substituted new materials.

Energy "supplied" through industrial energy efficiency comes in at a fraction of the cost of crude oil from offshore and Arctic reserves or even from lower-cost conventional

sources. Because energy efficiency helps cut the cost of manufacturing almost any product, regardless of what fuel is consumed, it enables industry to compete more effectively.

CANADIAN INDUSTRY PROGRAM FOR ENERGY CONSERVATION

The industrial sector accounts for 37 percent of total energy demand in Canada. That makes it our largest energy user. What's more, Canadian industry spends more than $10 billion annually to meet its energy requirements.

In May 1975, the Canadian Industry Program for Energy Conservation (CIPEC) was founded. Industry-administered and government-sponsored, CIPEC promotes and monitors energy efficiency throughout the Canadian manufacturing, mining and service industries. It has now developed into a

Energy efficiency lowers manufacturing costs and enables industries to remain competitive abroad.

50

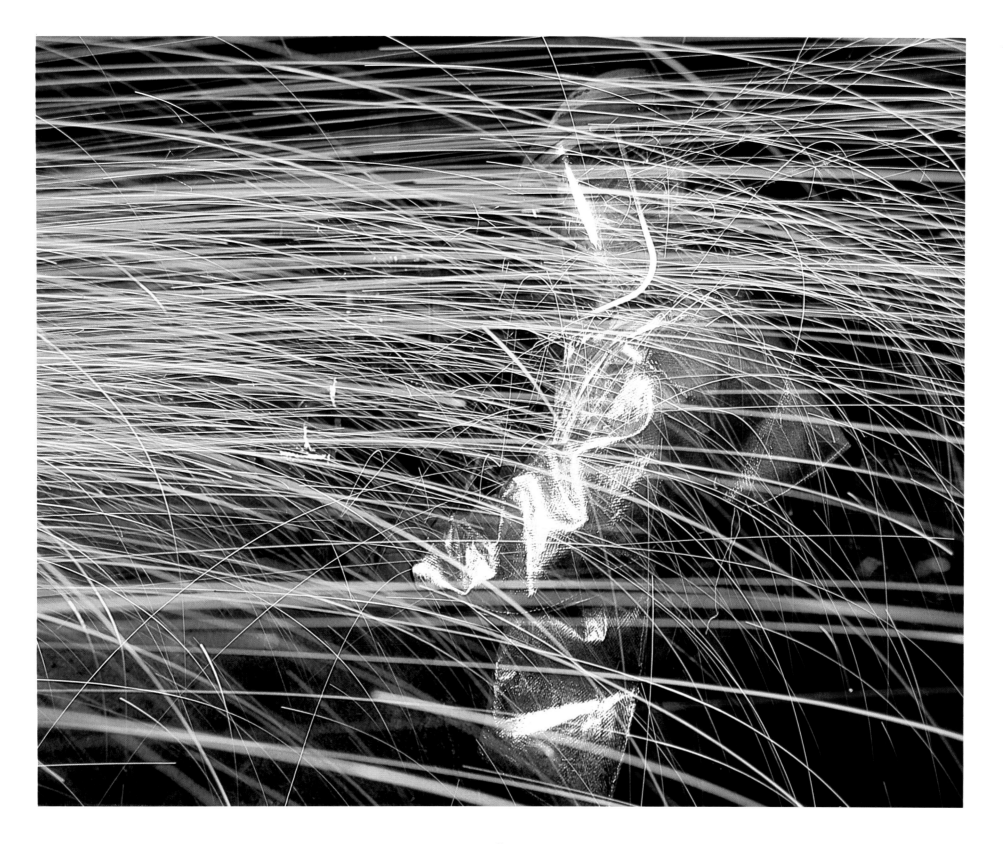

widespread cooperative network consisting of more than 650 participating companies represented by 14 task forces covering the chemical, electrical and electronic, food and beverage, mineral, mining, textile, and pulp and paper industries, among others.

Despite such achievements, Canada lags behind the performance of other industrialized OECD members. The OECD estimates that over the next 10 years Canadian industry has the economic potential to reduce its energy consumption a further 20 to 25 percent.

The potential, however, is not simply out there for the taking. It will require work and diligence on the part of industry and active encouragement from all levels of government. Success requires commitment, technological innovation and development — particularly where new production processes for energy-intensive industries are concerned.

As recent history has illustrated, any attempt to forecast the future availability and prices of energy resources is an uncertain business. Nevertheless, as a society, we may be prudent to insure ourselves against supply shortfalls and

The energy sector directly employs more than 300 000 people, representing almost three percent of total employment in Canada.

rapidly escalating energy prices. This task represents a
challenge today for Canadian industry to keep sound energy
management near the top of the corporate agenda. This
will not be an easy task, as many companies feel they
have already completed major energy-saving projects.

Today, the driving force behind energy management is
less conservation than improved productivity. In fact, in
some cases the energy savings are a small part of the
benefit. The larger gain is better international competitiveness, a cleaner environment and jobs for Canadians.

53

"...we must ensure that our efforts in economic development be carried out in an environmentally sensitive fashion. We must accept the fact that our industrial activities will not be worthwhile if in the end they spoil this planet to the point where life itself is in jeopardy."

Marcel Masse

TOWARD A CLEANER ENVIRONMENT

Industry's inventiveness in implementing energy management programs not only stimulates economic growth and creates employment, it also helps improve our standard of living and eases environmental pressures through reduced emissions and waste products.

Since the late 1960s, the public has become increasingly concerned about environmental issues. Canadians have sent a strong message to government and industry that environmental considerations should be an integral part of economic policy and planning. When industry incorporates environmental considerations at the design stage, society reaps the benefits of reduced clean-up and remedial costs. For their part, the corporations and industry associations that have decided that respect for the environment is in their own interest are recognizing that the cost of environmental protection is part of the responsibility of being in business.

Rather than developing what are often referred to as "end of pipe" or "stack" clean-up solutions, the Government of Canada actively contributes to the development of cleaner fuels, alternative energy sources, more efficient (and therefore less polluting) industrial processes and better combustion techniques with the goal of reducing emissions of harmful substances into our air, water and soil.

In 1975, fuel oil consumption represented 32 percent of the total energy used in the industrial sector. By 1987, energy management and substitution had reduced this share to less than 10 percent. In this period, industry saved 10 million cubic metres of oil and, in the process, avoided emitting millions of kilograms of sulphur dioxide into the atmosphere. At the same time, emissions of other pollutants such as carbon dioxide (the main cause of the "greenhouse" global warming trend) were reduced.

Of course, one way to lessen environmental damage is simply to use less energy, no matter what kind it is. Almost every kind of energy we use, or how we use it, has some impact on the environment. This is reason enough to maximize our efforts in energy efficiency. But the evidence also demonstrates that it is usually more cost-effective for industry to invest in energy management technologies than to purchase additional energy supplies.

IMPROVING CANADA'S COMPETITIVE EDGE

Because the Canadian industrial sector varies so widely, a
few examples have been chosen to illustrate what it is doing
to improve its competitive position.

Control room at a chemical plant: today the driving force behind energy management is improved productivity as much as it is conservation.

INDUSTRIAL PROCESS CHANGES

Instituting new or improved manufacturing processes is an obvious first step in upgrading efficiency. Innovations such as energy cascading (in which low-grade "waste" energy is reclaimed and, in some cases, upgraded for re-use), improvements in boiler efficiency, and reassessment of operational procedures have helped many companies become more competitive.

NEW, EFFICIENT AND VARIABLE-SPEED ELECTRIC MOTORS

The increasing use of automated equipment in Canada's mines has reduced production costs and increased productivity in this energy-intensive industry.

A large petrochemical company in Quebec replaced two large steam-condensing turbines with two 17-megawatt variable-speed electronic motor drives delivering 95 per-cent conversion efficiency. The estimated $11 million annual saving will pay back the cost of the project in less than two years.

ENERGY FROM WASTE

The trend of diminishing reliance on fossil fuels is especially true in the pulp and paper industry. Wood wastes generated by pulp and paper mills and neighbouring logging and sawmill operations now account for 67 percent of the total fuels burned by the industry.

The wood products sector has also lowered its conventional energy requirements through converting to wood wastes. Substantial improvements in the energy efficiency of dry kilns has largely been a result of conversions from natural gas-fired kilns to wood waste-fired energy systems. This permits an internally generated material to be substituted for purchased fuel.

Economic growth is sustained by a healthy, strong environment. It is important that government and industry take the lead in providing for economic growth and protecting the environment, but ultimately it is up to the people of Canada to decide how important clean air, clean soil and clean water are to the quality of their lives.

Wood residues can be burned to produce electricity, steam, or hot water.

Canada is the world's largest producer of hydroelectricity.

"Energy efficiency can only buy time for
the world to develop 'low-energy paths'
based on renewable sources which should
form the foundation of the global energy
structure during the 21st century."

(Report of the World Commission on
Environment and Development)

A NATION ON THE MOVE

Most of us are only vaguely aware of the quantity of energy we consume to heat our homes and run our factories. By contrast, we are all too conscious of the amount and cost of the gasoline it takes to fill the family car.

Transportation is the life-blood of our nation. The rapid movement of food and consumer goods across the country's great distances is vital to the health and well-being of all Canadians. The ability to travel as and when we want has become part of our modern way of life.

While many sectors of our national economy have been able to reduce their reliance on oil through energy efficiency and fuel substitution, the transportation sector still remains heavily dependent on refined petroleum products. Almost 99 percent of transportation energy comes from oil-based sources, while the transportation sector accounts for 63 percent of Canada's total consumption of petroleum fuels.

But gasoline, diesel and turbo fuels are not the only means by which we can power our cars, trucks, trains and airplanes. Why, then, is Canada so dependent on oil for its transportation needs?

Currently, more than 11 million passenger cars and three million trucks consume about 80 percent of Canada's transportation fuel. Surprisingly, perhaps, airlines use less than 10 percent, partly because the rising cost of aviation fuel sparked a switch to energy-efficient aircraft in the early 1980s. Railways use close to five percent of our transportation energy, as does the boating and shipping sector.

Fifty billion litres of crude oil — equivalent to 2 000 litres for every Canadian — must be refined each year to meet this country's demand for transportation energy. Because road vehicles are the largest single

consumers of petroleum products in this country, it is in this part of the transportation sector that the greatest gains in conservation may be realized.

During the oil crises of the 1970s, growth rates in transportation energy demand dropped from more than five percent a year, in the period preceding the first oil shock, to 2.4 percent during the late '70s.

But energy analysts now note that the rate of increased fuel use is moving closer to that of pre-OPEC days. High performance and larger automobiles are once again popular. Lower gasoline prices have reduced the motivation to investigate alternative fuel sources and have slowed the trend toward improvements in fuel efficiency. But further improvements are not only possible, they are easy to achieve during the normal product development of new vehicles.

The rapid movement of food, consumer goods and people across the country depends on readily available and affordable energy.

THE ROAD TO FUEL EFFICIENCY

In the transportation sector, even the smallest improvements in energy efficiency add up very quickly. For example, from 1978 to 1987, the average weight of the family car dropped to such an extent that fuel consumption was reduced by more than 27 percent. As a result, cars used only 8.3 litres of gasoline to travel 100 kilometres — about half the amount of fuel needed to drive the same distance in 1973.

Aerodynamics, or the streamlining of automobiles, has improved fuel efficiency by about 17 percent, while something as straightforward as reducing engine friction, for example, has improved fuel economy by four percent.

Similarly, turbochargers can decrease fuel consumption by 12 percent, while providing the same engine power. Though these devices were installed on fewer than half a percent of new cars in 1978, by 1987 this figure had climbed to more than four percent.

While much remains to be accomplished, research into

Today, there are more than 11 million cars on Canada's roads and highways.

67

electronic engine controls, low-profile radial tires, advanced combustion chambers and variable-speed transmission technologies could provide opportunities for further gains in fuel economy. Other measures, such as training in energy-efficient driving techniques, coordination of stoplights and observation of speed limits, are also helpful.

Most Canadians have experienced the frustration of cold-weather start-ups. But how many of us are aware of the increase in fuel consumption associated with winter driving? Starting the car and waiting for the engine to warm up can add 10 percent to our fuel bill. Waiting to warm up the interior can add another one to two percent. Installing snow tires, plowing through snow and slush and spinning wheels on icy roads all contribute to increased fuel consumption.

Fuel testing at an oil company's research laboratory.

One of the most significant trends in automotive history has been the switch to front-wheel drive. Although front-wheel drive saves only slightly more than two percent in fuel consumption, it allows vehicles to be downsized considerably, which translates into fuel savings. Still another fuel saver is the addition of an "overdrive" gear which can reduce fuel consumption by more than five percent.

In today's trucking industry, fuel efficiency is often part of staying competitive. Transport companies are reducing fuel consumption in vehicles of all sizes by using wind deflectors and other devices such as flarings and air dams, which reduce turbulence, and high-torque diesel engines, which produce more efficient low-speed operation. Automatic fan shut-offs and side-vented engine-cooling intakes can also improve fuel economy.

The matter of fuel consumption is more than mere economics — it has to do with the environment and our quality of life. Since cars and trucks account for 97 percent of atmospheric lead emissions in Canada, the Canadian government has announced a virtual phase-out of leaded fuels. Today, 87 percent of new gas-powered passenger cars run on unleaded fuel, compared with 46 percent in 1985.

Truck with wind deflector.

Traffic on the Pierre Laporte Bridge, Quebec City.

ALTERNATIVE FUELS:
THE WAY OF THE FUTURE?

Research is under way to develop alternative methods of powering cars and trucks that now use gasoline.

DIESEL FUEL

Although less than two percent of Canadian vehicles are currently powered by diesel fuel, this petroleum product ranks second after gasoline as a means of automotive propulsion. Since 1958, the demand for diesel-powered vehicles has risen sharply — largely because diesel engines provide 27 percent better fuel consumption. Research is continuing on ways to solve some of the minor irritants, such as noise and response time, that inhibit the greater penetration of this fuel into the marketplace.

NATURAL GAS

The tremendous advantage of natural gas as a motor vehicle fuel is that engines using it emit negligible amounts of carbon monoxide and sulphur dioxide and its combustion does not create smog. At the same time, there is no need for lead additives because natural gas has a higher octane rating than gasoline. In Canada, approximately 17 000 vehicles have converted to natural gas. The National Energy Board predicts demand will double by the turn of the century, as railroads, ferries and tractors convert to natural gas as a cost-effective substitute for petroleum-based products.

Because natural gas storage tanks add so much weight to a vehicle, present-day research is concentrated on lighter in-trunk fuel tanks, as well as methods to improve conversion equipment and promote safety.

PROPANE

Propane emits low levels of hydrocarbons and other pollutants and, like natural gas, needs no lead additives. Propane is easy to transport and cheaper than gasoline on an equivalent energy basis. Each year, Canada exports 40 percent of its propane production; the balance, some

71

2.8 million cubic metres, could be used to reduce gasoline consumption in this country by seven percent.

ALCOHOL FUELS

No energy source is as plentiful in Canada as the materials that can be used to make methanol and ethanol. These fuels can be produced from a variety of sources including natural gas, coal, wood and biomass.

Engines can be designed to run on pure methanol or ethanol and either product can be used as a fuel extender or octane enhancer. Henry Ford's Model T was designed to run on ethanol, and since World War II the product has been used as an additive in high-performance racing cars.

The benefits of alcohol fuels are numerous: they significantly reduce nitrous oxide emissions and generate no particulates; they can be used as environmentally safe octane enhancers to displace lead; they have a high octane number, which permits engines to be designed with higher compression ratios producing higher power outputs; and they can be stored and handled in liquid form.

Methanol, as a chemical, is already being shipped in large quantities across Canada and around the world. Only minor changes would be required to adapt the existing gasoline and diesel fuel production process to accommodate the addition of alcohols. A mixture of up to 10 percent ethanol or five percent methanol plus a co-solvent requires no change to a car's carburetor or fuel pump.

Research into other liquid fuels is also under way. Ammonia, vegetable oils and sewage sludge, for example, can all be used to produce liquid fuel for transportation. The possibility of transforming landfill wastes into vehicle fuel is another concept that holds tremendous potential both in reducing our dependence on oil and in addressing the ever-increasing problem of municipal garbage disposal.

ELECTRIC VEHICLES

In the 1920s, more than 50 Canadian cities had electric streetcars. By 1965, Toronto was the only one left. Canada had followed the lead of the United States in replacing the infrastructure of public transportation systems with roads built to accommodate privately owned gasoline-driven cars.

Cities were transformed as freeways and suburbs sprawled out from the core.

Today the benefits of electric vehicles are being considered again. For one thing, electric vehicles generate no pollution — either from exhaust, or from noise. For another, they have instant torque and their efficiency rating is 90 to 95 percent compared with the 25 to 30 percent rating for conventional engines. The cost of operating electric trolleys is, however, much higher per kilometre than diesel buses, partly because of maintenance of the overhead cables in the Canadian environment.

Light rapid train: Canada is considered a world leader in this "state-of-the-art" technology.

Batteries can also provide power for electric vehicles. As battery technology matures, electric fleet vehicles are becoming more practical. For applications in which driving distances are limited and the ability to re-charge batteries is assured, electric vehicles are already a viable option.

The disadvantages of electric vehicles include the considerable weight of the batteries (35 to 40 percent of total vehicle weight), the restricted driving range, and reduced performance in cold weather. Researchers hope to solve these problems by developing new batteries with higher power and storage capacity.

Electricity can also be used to provide power for trains. Already, a 130-kilometre line in the interior of British Columbia has been built to haul coal through a mountain pass. In addition, Canada has developed the LRT (Light Rapid Transit) system and is considered a world leader in this "above-ground subway" technology, which has considerable export potential.

Another technology that could give electric transport a boost is the fuel cell, a device in which fuel is converted directly into electricity. This technology has the potential to improve energy conversion efficiencies and, at the same time, reduce exhaust emissions. The immediate potential of fuel cells lies in submersibles such as exploratory submarines, in industrial applications such as cogeneration, and in large transportation vehicles including buses and locomotives.

Finally, there is solar electricity, the same technology that provides power for space stations. Photovoltaic or solar cells convert the sun's rays directly into electricity, which can be used immediately or stored in batteries for later use. A trans-Australia car race recently featured vehicles that were powered solely by photovoltaics.

*   *   *

Whatever fuel is used to power our vehicles, it must be plentiful, affordable, easy to handle and efficient.

Gasoline prices are currently low and supplies stable, but this situation will not last forever. Conventional oil-based non-renewable fuels are becoming less accessible and as they do, prices will increase. Petroleum can be expected to play a crucial role in the transportation sector for some time to come, but increasingly Canadians are recognizing that the movement of goods and people across this country consumes a large portion, perhaps too large a portion, of our oil-based supplies.

Research and technological innovation are helping reduce the cost of alternative fuel sources, making them more affordable and more available to both fuel companies and motorists. As the transition takes place it will generate new jobs, promote regional development and diversify economic activity. Because alternative transportation fuels burn more cleanly than gasoline, they will also help improve the quality of the air we breathe.

"Nations are formed and kept alive by the fact they have a plan for tomorrow."

José Ortega y Gasset (1883 – 1955)

BEYOND OUR BORDERS

Canada has prospered as a result of its energy resources. Energy exports last year brought in $12 billion, or 10 percent of our total export earnings. Over the last 20 years, Canadian companies have gained expertise in energy supply technology that can be merchandised abroad.

The changing international energy market will create new opportunities for Canadian exports. Although the United States currently buys the vast majority of the energy services we export, the industrialized countries of Europe represent a potential market for oil substitution technologies and certain energy management techniques. Finally, developing countries, with their need for conventional fuels and ways to improve energy efficiency, are an even greater and larger unexplored market.

CANADA'S ENERGY POWERHOUSE

Canada has been a net exporter of energy since 1969 and a net exporter of all energy commodities, including oil, since 1983. In 1986, of our total energy resource production, we exported 90 percent of our heavy oil and 20 percent of our light crude; 26 percent of our natural gas; 45 percent of our coal; nine percent of our electricity; and 78 percent of our uranium.

Energy has dominated world markets since the 1920s. During that time, oil, with its high energy value and easy transportability, has surpassed all other forms of energy to become the most heavily traded international commodity. Despite the fact that oil is a non-renewable resource and that prices and supplies are vulnerable to manipulation, and despite the fact that most countries are trying to find ways to diversify their energy supplies, oil indubitably

remains the most important and influential energy source in the world.

The discovery of oil at Leduc, Alberta, in 1947 brought Canada into the international market as a major exporter of petroleum. Between 1960 and 1973, oil exports as a percentage of domestic production rose from 23 to 66 percent. Since the Leduc discovery, estimates of Canada's supply potential have varied dramatically. It is now accepted that while the amount of conventional crude under the ground is much less than originally predicted, Canada's tar-sands and offshore potential, while more expensive to develop, assures this country of petroleum supplies far into the future.

Storage tanks for natural gas liquids, Edmonton.

Natural gas is not as easily transported as oil. Because it is six times as expensive as oil to distribute on an energy-equivalent basis, the United States remains, for the time being, our only export market for natural gas and we are its principal foreign supplier.

Distance constrains the viability of electricity exports as well, but technological innovations in the transmission of electricity have permitted significant growth in exports, particularly from Eastern Canada to the northeastern United States. In 1986, Canada exported almost eight percent of its generated power to the U.S., meeting 1.4 percent of that country's electricity demand.

The transportation of coal has always been a problem. Even though coal reserves are plentiful and Canada is an efficient producer, long distances are a major cost factor. Fifty percent of the cost of thermal coal lies in rail transportation from inland mines. Nevertheless, the export potential of coal may be improved by research into converting it into liquid fuels through co-processing with heavy oil and through the use of coal-generated steam in enhanced oil-recovery techniques.

Canada is also the largest producer and exporter of uranium, which is used to fuel nuclear reactors, in the western world. The world's largest producing uranium mine is in Saskatchewan. In addition, Canada has one of only five uranium refineries in the world. Canada has invested billions of dollars in developing nuclear technology. Although the CANDU reactor is well regarded around the world, the export market is small and highly competitive.

Darlington nuclear plant, Ontario. Canada has sold CANDU reactors to India, Taiwan, Pakistan, Argentina, South Korea and Romania.

NORTH/SOUTH:
AN ESSENTiAL PARTNERSHIP

The average North American consumes 330 times as much energy as the average Ethiopian. Unlike the industrialized world, where less than one third of the world's population consumes more than 80 percent of the world's resources and which must in the future strive to keep its energy demands in check, the developing world, if it is to achieve even a basic standard of living for its people, must find a way to secure greater access to the earth's supply of non-renewable resources.

Canada, with its wealth of expertise in most forms of energy, has already exported its know-how in large hydro-electric systems, oil exploration and distribution, and CANDU technology to many developing nations. Canadian consultants and contractors have been responsible for many dramatic hydroelectric projects — the first high arch dam (India's Idukki), the largest earth-filled dam (Pakistan's Tarbela), and the first extra-high-voltage transmission line (Tarbela to Lyallpur). We have helped Pakistan's oil and gas sector improve its management skills and increase production. In addition, we have sold nuclear reactors to India, Taiwan, Pakistan, Argentina, South Korea and Romania.

The export of conventional energy commodities and technologies is big business for Canada, but the other side of the coin is that the cost of these projects to developing nations is very great. Not only is financing difficult to obtain, but some major conventional energy projects have in the past had devastating effects on the environment. Today many developing countries have neither the financial resources to pursue projects of such magnitude, nor the desire to risk such damage to the environment.

International aid agencies no longer automatically assume that big is necessarily better. Some large hydro dams have been decommissioned within four years of construction because of silting. Other ecological problems related to changes in water current have created breeding grounds for disease. The construction of the Aswan Dam in Egypt, for example, ended the natural annual flooding that formerly fertilized the Nile's river banks. Since then, local farmers have had to spend a total of $6 million a year on fertilizer.

In some developing countries, families spend up to seven hours a day searching for firewood.

But the greatest change in export trading may result from the World Bank's recognition that money to build large energy projects is simply no longer available. Smaller projects are likely to become more common in the future.

ENERGY AND THE GLOBAL HABITAT

In Canada, energy production and consumption generate 85 percent of all nitrogen oxide pollutants, 65 percent of lead emissions, 41 percent of sulphur oxides, 23 percent of mercury contamination and 18 percent of particulates. This assault on the environment is also generated by other countries with similar energy use patterns.

If developing countries were to follow the example of the industrialized world regarding energy use, the environmental consequences would be disastrous. Yet as the Brundtland Commission has so clearly stated, developing nations are in desperate need of increased access to energy in order to achieve economic growth. Perhaps the greatest challenge facing the world as we enter the next century will be to find ways to meet the energy demands of the poor without repeating the mistakes of the rich.

By keeping environmental considerations in the forefront of international decision making, industrialized and developing countries can work together to better achieve their common energy objectives. The potential for more environmentally sensitive, renewable energy technologies has barely been investigated.

The average North American consumes 330 times as much energy as a person living in poorer developing nations.

RENEWABLE ENERGY:
SOLUTIONS FOR TOMORROW

What developing countries lack in financial wealth they make up for in their abundance of renewable energy resources. Indeed, developing countries and the agencies that lend to them are beginning to examine more closely smaller, decentralized projects that make use of technologies such as wind and solar energy, bioenergy, small hydro and cogeneration. While low oil prices have diminished the short-term prospects of some of these technologies in Canada, most remain competitive in foreign markets where conventional energy supplies are still unacceptably expensive or simply unavailable.

Canada has already demonstrated its technological capability in many renewable energy technologies both at home and abroad. Our reputation for rugged, reliable solar heating systems has been aptly illustrated in the Florida fruit belt. When temperatures in the sunshine state dropped below freezing, U.S. systems froze, while Canadian

Hydro power, second to wood among the renewables, has been expanding at nearly four percent annually, but the remaining potential for this form of energy is huge.

products continued performing. Many of the world's largest solar water heating and solar air heating systems were manufactured in Canada for installation in Europe and the United States.

Canada's solar energy industry also exports solar crop dryers, which can supplement or even replace diesel oil systems and can help prevent major crop losses that occur when food cannot be dried quickly enough after harvest.

Canadians have also developed expertise in small hydro technology — technology that is working at home and in developing countries. In Canada, as the financial and environmental risks of large hydro projects have risen, some

Photovoltaic systems can be used in almost all developing countries to provide water pumping and water purification.

private entrepreneurs have taken a keen interest in reopening former small hydro sites with existing dams and selling the electricity they produce to the local utility.

Our expertise in bioenergy technologies could also be put to good use in developing nations. In many of these countries, the decimation of firewood supplies is reaching crisis proportions. Many families spend hours each day searching for firewood. Some spend as much as one third of their family income on fuel.

This country has an established reputation in wood gasification and waste combustion technologies. Our knowledge of bioenergy technologies will also help us participate in the growing world market for ethanol, which can be used as an additive to or replacement for oil-based transportation fuels.

If the shortage of firewood in developing countries is a pressing problem, then so too is the need for small, rugged electrical systems for remote applications. Less than 25 percent of the developing world has regular electrical service — a figure that drops to less than four percent in rural Africa. This opens a major market niche for solar electric (or photovoltaic) and wind energy systems.

Where villages are far apart, population is low, or consumption is limited, conventional electrical networks are prohibitively expensive and impractical. Simple, reliable and durable photovoltaic systems convert the sun's rays directly into electricity. They can be used almost anywhere to, for example, refrigerate vaccines and provide water pumping and water purification. While the cost of photovoltaic cells is still high, they are often the most economical, and sometimes the only practical, power source for such applications. Through technology transfer agreements, an Ottawa-based photovoltaic company has exported solar cell manufacturing facilities to both India and China.

Canada has a reputation worldwide for rugged, reliable solar heating systems.

Where wind regimes are suitable, small wind systems may be another option. Canada's wind energy industry is perfecting wind machines for both electrical generation and water pumping. A Toronto-area firm has recently installed a 50-kilowatt vertical-axis wind electric machine on the Turks and Caicos Islands off the eastern tip of Cuba and hopes to install additional, larger machines in the area.

While Canadian companies have won praise for the development of new energy supply technologies, they have also made their mark in the area of energy efficiency. Innovations such as R-2000 building construction, high-performance windows, the "Light Pipe (TM)", and "energy audit" programs, which have been especially popular in Europe, show major export potential to both industrialized and developing nations.

Wind is a safe, clean and renewable energy source.

THE ROLE OF GOVERNMENT

Governments in Canada play a role as well in the export of energy and related technologies. At the United Nations Conference on New and Renewable Sources of Energy, held in Nairobi, Kenya, in 1981, Canada agreed that new energy sources would be a priority area for international assistance.

Some of the avenues open to export entrepreneurs, such as the Canadian International Development Agency (CIDA) Business Cooperation Branch and External Affairs' Program for Export Market Development (PEMD), are

Canadian engineers working in Pakistan: our expertise in hydroelectric projects is one of our greatest export commodities.

95

already well established and open to proposals that support technology transfer from Canada to developing countries. In addition, Petro-Canada International provides assistance to less developed countries in the exploration of hydrocarbon resources.

Investment in research is also making it possible for Canadian entrepreneurs to make inroads in new technological directions. For developed and developing countries alike, much of this technology will offer new hope in the search for creative, environmentally sensitive ways to meet their energy requirements.

*   *   *

In the report of the World Commission on Environment and Development, committee chairperson Gro Brundtland noted that any improvement in world economies crucially depends on increasing the amounts of energy derived from sources that are dependable, safe and environmentally sound.

The Brundtland Commission believed that we can act together to establish a sustainable world economy, one in which increased energy use and economic growth do not further imperil our environment. Over the next 20 years, the world will be forced to make major decisions in this regard and will face enormous tasks.

It was only 300 years ago that Newton demonstrated the relationship between energy and mechanical power, and changed our world forever with his observation. In choosing our future path, let us hope that concerted, cooperative action will prevail over conflicting interests and narrow thinking. Only a failure of imagination and courage can stand in the way of an environmentally sustainable energy future. At stake, after all, is the planet.

"Hope is itself a species of happiness and perhaps the chief happiness which this world affords." Samuel Johnson (1709 – 1784)

"To act without rapacity, to use knowledge with wisdom, to respect interdependence, to operate without hubris and greed, are not simply moral imperatives. They are an accurate scientific description of the means of survival."

Barbara Ward (1914 – 1981)

ACKNOWLEDGEMENTS

The support of all those involved in the production of ENERGY: THE POWER OF CANADA has been invaluable. They include Energy, Mines and Resources Canada (EMR), Supply and Services Canada (SSC) (especially the Client Services Group and the Canadian Government Publishing Centre), the design group Addventures, and Passmore Associates International.

Those who deserve special thanks for their excellent contributions are Brent Moore of SSC, who administered and coordinated the project so well; Debora Morton, Dave McGoldrick, Pierre St. Jacques and Neil McEachern of Addventures, for the quality of their design work, carried out with patience and constant good humour; Cécile Suchal of EMR, for her determination and ability to meet apparently impossible deadlines; editors Patricia Finlay for her linguistic dexterity and judgement and Ruth Tabacnik for her attention to detail; and finally Nicola Fletcher for her guidance and hard work in bringing together the latter stages of the project.

PHOTOGRAPHIC CREDITS